Lake Sacajawea

Longview's Treasure

Published by Lake Publishing Company
2938 Laurel Road
Longview, WA 98632

Illustrations: Rick Cavens
Map: Rick Cavens
Watercolor painting: Sharon Pedersen

COVER PHOTO: Lake Sacajawea in the fall.

HISTORICAL PHOTO CREDITS: *The Daily News, Longview, Washington*, pages 40, 58.;
Longview Public Library, Longview Room Collection, pages 12, 19, 21, 22, 25, 29, 30, 31, 89;

Library of Congress Catalog Card Number: 97-74061
Cavens, Travis R. , 1935-
Lake Sacajawea : Longview's Treasure / text and photography by Travis R. Cavens
ISBN 0-9659385-0-6
1. Longview WA. 2. R. A. Long

Printed in Korea

➤ *Gazebo on Lions Island.*

LAKE SACAJAWEA

WE ARE KNOWN BY OUR LAKE

New visitors to Longview are usually taken by friends or relatives to Lake Sacajawea to show them its beauty and peacefulness. And of all the things that they see in our city, it is their images of the Lake that they most often retain.

When I was in Zaire during the Rwanda refugee crisis in 1994, I met a Seventh Day Adventist missionary who had been working in East Africa for decades. He was a Dutchman with the unwieldy name of Dete Vanderwerff. He asked me, "Travees, vere do you come ffromm?" I told him that I lived in a small town called Longview, Washington.

His very next words were, "Ah Longview" - pause - "Lake Sacca-jah-veea." I sat astounded. He not only had been here, but his first recall of our city was that of the Lake.

---- Travis Cavens

Dete Vanderwerff.

◄ *The Lake with summer colors.*

Acknowledgments

The author is grateful to his son Derek, who passed on his commercial photography skills to a member of the older generation, and to his wife, Phyllis, who was and always has been so supportive.

He is also thankful to Sue Maxey of the Longview Public Library, Richard Bemm and Al George of the Longview Parks Department, Ted Natt of The Daily News, Reverend Harlan Gilliland of Longview Community Church, Bob Carpenter of Consolidated Diking District No. 1, Charles and Danielle Craig, Roger Grummel, and the Diamond Jubilee Committee.

This book could not have been completed, however, without the extraordinary work of John M. McClelland Jr. who wrote the definitive history of the city, "R. A. Long's Planned City, the Story of Longview," 1976, Longview Publishing Company.

Logo designed by the Diamond Jubilee Steering Committee to commemorate Longview's 75 years of existence. Their theme: "Celebrating the past, Fulfilling the Vision"

◄ *Aerial view of the Lake during the Fourth of July.*

◄ *(Previous Page:) A summer view from the Washington Way bridge.*

THE LAKE AS LONGVIEW'S TREASURE

From the air, Lake Sacajawea shows her gentle curve and the dense green of trees packed around her borders. There is an occasional glimpse of the three-and-half miles of pathways, used day and night, rain or shine, by runners and walkers alike. Crowds of people from all over Southwest Washington are seen basking in the sun, in the music, and in the excitement of human activity. This is the gathering place.

The Lake is a beautiful gem, to be treasured and preserved, but it didn't start out that way. It originated as a brushy, mucky slough.

A metamorphosis began in 1918 when Mr. Robert Alexander Long and the men of his lumber company came to the area to build the world's biggest sawmill and along with it a planned city. Because the area was part of the floodplain of the Cowlitz and Columbia rivers, they found they needed fill dirt to make a huge dike to protect the land they were going to build on, and so they dredged the slough. Thus the Lake was born.

As their vision grew, they saw the Lake not only as a magnet to sell homes, but as a grace note of tranquillity and beauty. It was to be a place where citizens would gather to relax and to forget the rigors of work in the mill. And so it was for a time.

But the worldwide Depression nearly killed the dream. The Lake was not kept up, and it was almost sold to be lost forever for public use. Only in the last few decades has her beauty been restored and have the people begun once again to use it as a gathering place.

15TH
NEW WESTSIDE
OLYMPIA WAY
WASHINGTON WAY
OLD WESTSIDE
KESSLER
NICHOLS
OCEAN BEACH
LOUISIANA
OLYMPIC ADDITION

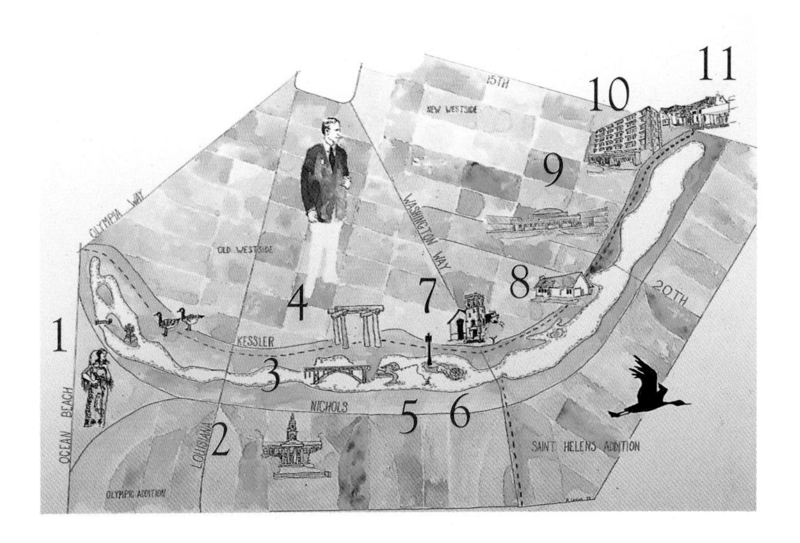

KEY TO MAP

1. *Japanese gardens*
2. *R. A. Long High School*
3. *Footbridge*
4. *Double arches*
5. *Martin's Dock*
6. *Lions Island*
7. *Longview Community Church*
8. *Elks War Memorial*
9. *Kessler Elementary School*
10. *St. John Medical Center*
11. YMCA

MR. R. A. LONG — LUMBER BARON

R. A. Long, *Longview's founder.*
Longview Public Library photo.

R. A. Long was rich in the early 1920s, with personal assets worth $20,000,000. He had a valet who laid out his clothes and shaved him every day, plus he often traveled across country in his own private railroad car.

Mr. Long had made his money by starting a lumber yard in 1876 with Victor Bell, whom he subsequently bought out. In 1918, the Long-Bell Lumber Company had grown so big that its corporate office was housed in the tallest office building in Kansas City, Missouri, the 15-story, R. A. Long Building.

But he was in debt when he died at age 84 in 1934. The Depression had hurt his lumber business, but in addition to that, Mr. Long gave away over $1.7 million of his own money to the city and the dream he had founded. Much of it was spent on the Lake and its environs with $750,000 going for R. A. Long High School, $170,000 on Lake Sacajawea Park, $5,695 to Kessler School grounds, and $6,235 on the sunken gardens in front of the hospital. He donated $25,000 to start the fund raising campaign to build Longview Community Church.

Many criticized Mr. Long's excessive spending, but he defended his generosity by declaring, "Now some have said that this is satisfying an old man's dream. Well, I am old in years. I guess that's right. It may be the dream of an old man, but I am willing to stand on my record and let you determine whether or not expenditures of this kind will add to a community."

➤ *Dan Dowling peacefully paddles his kayak on the Lake built by R. A. Long's company.*

R. A. LONG HIGH SCHOOL — A GIFT

Longview's first high school was not built with tax dollars, but was instead a personal gift from lumber baron R. A. Long. He spent $750,000 of his own money. Most people at that time assumed that Mr. Long was still so very rich that he could easily pay for it, but this benevolent act turned out to be a financial sacrifice for him and one that caused a great deal of bitterness among his family members.

By 1928, his fortunes were being drained; therefore, he was forced to mortgage his office building in Kansas City to obtain the money for the school. The Kansas City building was 15 stories high, the tallest building in the city at that time, and the family counted on it for their security against hard times. They were very upset and bitter towards Longview when they learned of Mr. Long's action.

The architectural design of the high school is classic colonial with tall columns, topped off with a clock tower. It is similar in style to the Longview Library and the Monticello Hotel, but it is even more imposing because of its location. It is framed by a vast expanse of land with tree-lined boulevards, fronting the Lake.

But it was fate that gave it that setting. The school was built away from the Lake on property that wasn't as valuable as lakeside property. The two blocks in front of the school were to be divided up into lots to build houses on. But following the opening of R. A. Long High School in 1928, economic hard times hit, and those two blocks never sold, leaving instead a panoramic view of the school.

R. A. Long High School logo.

➤ *R. A. Long High School.*

COMMUNITY CHURCH

As the construction of Longview and the mill commenced in 1922, the population mushroomed. It soon became apparent that there was no church to serve the spiritual needs of all the new people. Many residents wrote to R. A. Long urging the formation of a single interdenominational church. This led to a group of Methodists organizing the Community Church, which held its first service on October 21, 1923, in Long Bell's social service center, the Community House.

There was a young Methodist minister, Edwin H. Gebert, who had been riding the train from Tacoma down to Longview to preach in the construction camp. He was asked to lead this new church and continued to do so for 35 years.

But Reverend Gebert didn't like holding services in the auditorium of the Community House. He wanted a real church building and wrote to R. A. Long about it. Once again, Mr. Long opened his purse to donate the first $25,000 towards the $89,500 needed for construction.

In the towers of the church are the chimes that cost $7,200 and have 22 bells, ranging in size from the small 475 pound bell to the huge 2000-pounder. They are an additional gift from Mr. Long.

The Ku Klux Klan requested in 1925 that Reverend Gebert give the opening prayer for their June 14th meeting at the ballpark. The Board of Directors of Community Church voted this down, giving the reason that it would "interfere with our church services."

➤ *Longview Community Church.*

Community Church
THE TRANSFORMATION

WHY THE LAKE WAS BUILT

Lake Sacajawea is not a natural lake. It was actually built by one of the biggest lumber companies of the time, Long-Bell Lumber Company with headquarters in Kansas City, Missouri.

In 1918, the company realized that most of the usable timber of the south had been cut. It was decided, therefore, to move west to the virgin forests of Southwest Washington and there construct the largest lumber mill in the world. But a city to go with it was not part of the original idea. Instead, Long-Bell planned to buy the mill site and just enough land around it upon which to build shanties for the workers, which was standard practice of the times.

However there existed one major problem related to the flat farmland that was chosen for the mill. The entire area was a flood plain. When the Cowlitz and the mighty Columbia rivers were high, steamboats sometimes passed over the spot that is now downtown Longview. So the executives of Long-Bell argued over various ways to prevent flooding of their mill and financial disaster.

With the persistent urging of the company's chief engineer, Wesley Vandercook, it was finally decided to buy up the entire valley floor and put a 13-mile dike around it. In the middle of all this land, there was a mucky, brushy, mile-and-half strip of water called Fowler's Slough, which at one time had probably been a branch of the Cowlitz River. It was brilliantly reasoned that this useless slough could be dredged and that the earth from it could be used to build the dike.

Nine teams of horses are seen pulling scrapers to shape Lake Sacajawea.
Longview Public Library photo.

But it was a costly undertaking. Long-Bell eventually spent $2.6 million to buy the land and $3.2 million to build the dike and drainage canals. Early on, R. A. Long realized that he needed some way to recoup this immense investment, so he turned to a nationally known land developer, J. C. Nichols, for help. Nichols advised him to plan an entire city and to divide the land into lots to be sold for businesses and homes.

They hired an eminent landscape and city architect, George Kessler, to begin drawing the plans of the city. Those plans called for Fowler's Slough to be transformed into a lake that would be surrounded by a park and fine homes. A Seattle nurseryman, John Null, was given the job of park superintendent, a post he held for 49 years. He directed the landscaping and tree planting to make Lake Sacajawea a beautiful centerpiece for the selling of the surrounding real estate.

This painting of Sacajawea by Joe Knowles hangs among the murals in the lobby of Longview's Monticello Hotel.
Longview Public Library photo.

NAMING THE LAKE AFTER A LEGEND

Lake Sacajawea started out as a muddy, brush-choked branch of the Cowlitz River called Fowler's Slough. As the Lake began to take form and people could see the beauty that it promised, it was felt that it should have a new name. R. A. Long asked the fledgling newspaper of the town, the Longview Daily News, to conduct a contest.

The winning name was Sacajawea, after the Shoshone young mother who had accompanied the explorers, Meriwether Lewis and William Clark, as they paddled their canoes down the Columbia River in 1805 on their way to the Pacific Ocean. Lewis records in his journal that they stopped at the entrance of the Cowlitz River to talk to the Cowlitz Indians, but there is no mention of Sacajawea at that meeting.

The Lake, therefore, is not named after her for something special she did in the Longview area, but rather because of her legendary fame. More memorials honor her than any other American woman.

Sacajawea was probably 16 years old and had just given birth to a baby boy when the expedition left North Dakota. She was able to find food for the group and to make friends with Indians on the way.

When they were near the Yellowstone River, one of the boats almost capsized. Her husband, Toussaint Charbonneau, was frozen with fear. Lewis wrote : "The Indian woman, to whom I ascribe equal fortitude and resolution with any person on board at the time of the accident, caught and preserved most of the light articles which were washed overboard." She did this while balancing her baby on her back.

The arches of 1973 at Lake Sacajawea Park.

The original arches of 1924.
Longview Public Library photo.

THE DOUBLE LOG ARCHES

Beckoning users of the Lake are the double log arches at the west end of Hemlock Street. The massive cross timbers are 20 feet long and rest on support logs that rise up 14 feet. The individual logs are each two feet in diameter.

The arches didn't exist when the Lake was created, but were added decades later at the 50th anniversary celebration of Longview. Ron Johnson, a Longview resident, was selling placemats that pictured the original arches used in the Pageant of Progress in 1924 when he had the idea to build a duplicate arch. He managed to persuade local businesses to donate the architectural design, the logs, and the labor.

During the dedication ceremony on June 28, 1973, antique Model A and Model T Fords drove through the 16 foot "tunnel."

As impressive as the arches are, they are not as big as the original arches built in 1924, which were made of old growth Douglas firs. There were two arches back then, located at the east and west entrances to the city. One of them was at the intersection of Washington Way and Vandercook Way, near the West Kelso border.

A string of cars and 111 parade floats passed through the arches on July 31, 1924, to open the four-day celebration of the dedication of the new Long-Bell lumber mill. The festivities included a concert by the Royal Rosaria band from Portland, airplane trips, visits to a navy destroyer, an aerial circus with a wingwalker, and evangelist, Billy Sunday.

The YMCA is seen on the right, while on the left is Memorial Hospital, which later become St. John's Hospital.
Longview Public Library photo.

COMPANY SOCIAL CENTER TO YMCA

The YMCA today is an important part of the life of many residents of Longview, whether they are swimming, camping, or weight training. Runners often bolt from its doors to complete a run around the Lake.

But it wasn't always a "Y." It started originally as Long-Bell's company-owned "Community House." It contained a pool parlor, confectionery, church, movie house and lounge. Its purpose, furthermore, was to help with the personal problems of its employees.

The first director was incompetent and was soon replaced by Charles Nutter, a charming man from New England who had worked for the YMCA. By 1924, he had persuaded Long-Bell to deed the Community House to the YMCA, arguing that the company should put aside anything that looked like paternalism and choose the "Y" as "an independent, stand on its own legs' organization in which workmen themselves have a voice through committee services."

YMCA.

THE HOMES AROUND THE LAKE

At the beginning of Longview, the Lake formed a boundary between the richer people of the fledgling city and the working class. It turned out that way because J. C. Nichols, Mr. Long's land developer friend who did so much to beautify Kansas City, argued successfully that Longview must be planned in order to provide harmonious living between the industries, commercial section, and residential areas.

The plans called for the homes of thousands of millworkers to be close to the lumber mill, since few people could afford cars and would be walking to work. The first place chosen for them was called the St. Helens Addition, which stretches from Nichols Boulevard along the southwest portion of the Lake and south to the industrial area.

A later housing development, the Olympic Addition, curves around the northwest border of the lake. In both additions, St. Helens and Olympic, the streets were made of gravel.

Across the Lake, however, the streets were paved with diamond-shaped squares of concrete in order to sell the lots to richer buyers. Even though this section is on the east side of the Lake, it was called the West Side Addition, since it fans out west of the Monticello Hotel, between two streets, Washington Way and Olympia Way.

There was a triangular section of land between Community Church and Kessler Grade School that was developed in later years and was dubbed the New West Side. This led people to refer to the original West Side Addition as the Old West Side.

Many beautiful homes in the Old West Side were designed by architect, Arch Norman Torbitt. They range in style from Spanish stucco to Colonial to Tudor.

Tudor houses are characterized by their narrow vertical windows framed by half-timbered peaks above the front door and gables. Characteristic of this is the home he designed for S. Mark Morris, who was an executive for Long-Bell. In 1957 Mr. Morris, age 80, attended the dedication of the high school named after him.

Mark Morris home in 1924 at 1433 Kessler Blvd.
Longview Public Library photo.

➤ This painting by Longview artist Sharon Pederson is of the home owned by Mr. and Mrs. Robert Kirchner. It was built in 1932 for a Long-Bell executive. The huge wooden beams framing the walls are pieced together without nails. There are 14 rooms and 5 fireplaces, constructed with trade skills now forgotten.

Old West Side homes along Kessler Boulevard.

SHARON PEDERSEN ©

THE LAKE NEARLY SOLD

As the worldwide Depression developed in the late 1920s, construction throughout the United States slowed, and, with it, the need for lumber. Long-Bell began to have serious financial problems, forcing the company to lay off many of its workers and to drop support of civic institutions such as the library, the YMCA, and the Chamber of Commerce.

In addition it was no longer able to pay for the maintenance of its beautiful parks, and so they began to wither.

On April 23, 1938, an even more unthinkable thing happened — Lake Sacajawea Park was put up for auction. Long-Bell Lumber Company couldn't pay the back taxes owed on it, so the Park was to be sold to the highest bidder to be used for private purposes. One entrepreneur actually distributed a map showing the land divided up into 50-foot parcels extending down to the water's edge.

Fortunately, Lake Sacajawea Park, plus the Civic Center park, the sunken gardens and the Highlands playground were saved because Doctor J. L. Norris, a general practitioner, and Dr. R. S. Howell, a dentist, led a rescue operation to wheedle enough money to satisfy the committee of bondholders set up to oversee Long-Bell's reorganization.

In their honor, the bondholders said, "It is our hope that the citizens of the community will really appreciate these contributions and that the future generations will realize the foresight of your efforts in preserving them and maintaining them."

Kessler Boulevard and Lake Sacajawea in 1926. Longview Public Library photo.

PLAYTIME AT THE LAKE

Once upon a time, the Lake was clear. People swam and played in the waters. There were water carnivals with swimming races and decorated floats.

In 1928, the first loggers' version of a rodeo attracted thousands of people to the Lake bank. There was log chopping and bucking, ax throwing, plus the most popular event, log rolling, which inspired the name of the event, "Rolleo." Later Rolleos featured huge poles of Douglas fir erected in the Lake for "tree" climbing contests.

But the beauty of the Lake began to fade after a few years. Rainwater from the surrounding hills washed rich soil into the Lake, turning it murky and promoting the heavy growth of weeds.

The Depression dried up maintenance funds. The grass and shrubs grew so high that people feared to use the Park. The lawns turned brown when the irrigation pump broke and wasn't replaced. Mothers forbid their children to swim in the Lake because of the possibility of getting polio. And so the Lake park became something to look at, but not play in.

But times have improved in recent decades. Money is available to water and mow the lawns and trim the shrubs. The jogging phenomenon has spurred the building of trails and lights, and so now, day or night, people are using the park.

Once again Lake Sacajawea, has become the gathering place where people play and relax.

Swans have been introduced to the Lake several times over the years, but have not survived for very long.

Photos from Longview Room Collection of the Longview Public Library.

➤ *The American Legion bathing beauty contest of 1926 was won by Miss Kelso, Lucille Oatman, pictured on the far right.*

⮟ *Crowds gathered in 1927 for water festivities on the Lake.*

Summer Action

Looking for bargains at the flea market.

➤ *(Previous Page:) Log rolling contestants at the Timber Carnival on the Fourth of July.*

THE FOURTH OF JULY — WOW!

Each year, Lake Sacajawea is host to the biggest show in town, the Fourth of July celebration. Crowds of 40,000 gasp in unison as over a thousand rockets burst overhead during the nighttime fireworks extravaganza.

But many families arrive much earlier in the day to spread their blankets on the grass and listen to the live bands. They may wolf down a Lions Club breakfast and then later munch on an Altrusa elephant ear or some other delectable from one of the 18 food booths.

Many will stroll over the foot bridge to paw over the items for sale among the 100 flea market vendors. They may cheer on their children in various foot races or watch a semi pro football game.

Several thousand will gather at the Timber Carnival show to be thrilled by ax throwing, log rolling and "hot saw" bucking. They will squeal as pole climbers ring the bell high above the crowd and appear to almost free fall to the ground, taking giant strides with their spurs.

All this happens only because citizen volunteers have taken on the responsibility to see that it happens. They are the members of the Go 4th Committee who gather funds to pay for the fireworks display, the lumberjacks show and a myriad of other expenses. They sell Go 4th lapel buttons and obtain support from local industries.

The president of the Go 4th Committee was Bill Rowlson, who held that volunteer position for 17 years. It was his energy and enthusiasm that kept the event alive and growing.

An early breakfast is served by the Lions Club.

▲ *Carefully groomed collector automobiles at the Car Show.*

➤ *Young majorettes in formation.*

◀ *The Fourth of July parade moves down Kessler Boulevard.*

▲ *A straining member of the Lumberjacks Association has traveled to Longview to compete in the log bucking contest.*

➤ *A line of food venders, representing many of the area service clubs, sell hot dogs, elephant ears, cold drinks and more.*

◄ *Two golf balls splash near the pickup boat in the Longview Firefighters floating golf contest.*

LONGVIEW BRIDGE IS FALLING DOWN

One Fourth of July celebration made news across the nation. The fireworks display on Lions Island had just ended in 1968 when the wooden foot bridge at Hemlock Street collapsed, dumping 100 to 200 people into the night waters.

Bystanders dove in to rescue those clinging to pieces of wood. Tow trucks arrived to lift sections of the fallen bridge so rescuers could find those that might be trapped, while scuba divers searched in the pitch black water. No one was killed, but 80 victims were taken to hospitals. A steel bridge now crosses the Lake.

▲ *Rescue operations on July 4, 1968.* The Daily News Photo

➤ *A thousand shells rise from Lions Island, lighting the night sk*

▼ *The new aluminum footbridge.*

◄ *Water lilies line the Lake.*

▲ *Open air art show and craft show on a July Fourth.*

On a warm summer evening the Lower Columbia Community Band soothes the souls of rapt listeners.

Runners Sonja Cavens and Jamie Hoffman streak the 3.5 miles around the Park.

ELKS WAR MEMORIAL

Cities since the time of Roman conquest have erected statues of their great war heroes. In the 1940s the United States was flush with victory pride after World War II, and it would have seemed appropriate to erect a similar monument in Longview.

But that was not the case. Instead, the Elks Club of Longview built a war memorial that was different. They constructed a building that was dedicated to children at play. It is a wooden framed structure with peaked roof, snugly situated on Kessler Boulevard. Inside there is a fading wooden plaque, high on the far wall, that simply reads:

LIVING WAR MEMORIAL
B.P.O.E. 1514 JUNE 1947
TO THE YOUTH OF AMERICA

On the opening day, there were outdoor swings, and teeter-totters and a slide so high that it inevitably brought thrilling fear to those who ventured on it. The Elks continued operating the play center each summer until 1975, when they transferred ownership to the City of Longview.

Now the Parks and Recreation Department opens it to the community for activities such as nature day camps and art lessons.

In 1948, the American Red Cross commandeered the building, using it 24 hours a day for two weeks, as the community fought the rising flood waters of the Columbia and Cowlitz rivers that threatened to breach the dikes.

➤ *Day campers age 5 to 7 begin exploration as part of a Parks and Recreation Department program.*

Elks War Memorial
City of Longview
Parks & Recreation
2121
KESSLER BLVD
Art Room
Multi-Purpose
Room

➤ *Kyle burns one in to his dad.*

➤ *The Elks War Memorial playground, one of three playgrounds in the Park.*

➤ *Visitors enjoy a picnic in the park.*

Fall Repose

North end of the Lake at dawn.

South end of the Lake.

Long distance runners in a Southwest Washington interschool meet.

Winter Struggle

The Lake froze over in the winter of 1956-57, allowing skaters on the ice. The Daily News photo.

A boy tosses bread to the seagulls standing on the frozen lake fringes in January of 1996.

FLOOD INSURANCE

Longview, which was built on a flood plain, is threatened by flooding of the Columbia and Cowlitz rivers or by torrents of rain rushing off the surrounding foothills. Realizing this danger, Long-Bell in 1923 constructed a dike, 13 miles long, plus a series of drainage canals. These flood barriers channel water out to Coal Creek, where huge pumps lift excess water from the canals into Coal Creek Slough and pass it safely on to the Columbia River.

The initial design of the Lake created an unforeseen problem. Each year the heavy rains brought in silt that muddied the waters and fostered the growth of plants that began to choke the Lake. In 1979, there was a $5 million effort to correct the problem by building an eight-foot-wide culvert under the east bank, paralleling the entire length of Kessler Boulevard. The muddy rain waters are now blocked by a large gate and diverted through this giant tunnel.

If the Cowlitz River is fairly clear, the gate is opened so that fresh water can be pumped from the river to flush the Lake. The Lake is much clearer, but it still gets muddy, primarily in its middle section. It is suspected that there may be hidden springs that bring up silt.

Downtown Longview is given added flood insurance by the Lake itself. In the event of a severe flood, the same north-end gate can be opened to let the excess waters flow into the Lake. Since it extends over 50 acres in its mile-and-a-half length, 16 million gallons of water can be absorbed for each one foot rise in the water level.

➤ *Out for a brisk wintry stroll are a dog and his man.*

CHRISTMAS GLOW

Christmas doesn't seem to really begin in Longview until those radiant lights strung fancifully among the trees of Lions Island flash on, and Santa is seen guiding his flying reindeer over the Lake waters. From early December through mid-January, Frosty the Snowman waves a holiday welcome amidst the intense red and yellow lights that shimmer in the Lake waters.

This herald of Christmas is a work of love by the members of the Lions Club, who each year string some 15,000 lights, putting in over 200 man-hours to do the job.

But the Yuletide delight wasn't always so. Before 1964, there was only a neglected island in the middle of the Lake with thick underbrush rising up eight feet. It was "Mr. Lion," Ernie Kuntz, who shared his vision with his fellow club members. They then went to work, clearing the land and planting shrubs. Concrete bulkheads were constructed, and a new wooden foot bridge was built.

In recent years a transformer was installed on the island that doubled the electrical capacity, so that each Christmas season the amount of power consumed is equal to heating and lighting a home for more than seven months.

There are four islands in Lake Sacajawea. Before 1969, none of them had a name. But that year, the city of Longview officially named this one island "Lions Island," in honor of the club that had done so much to share the excitement of Christmas.

➤ *The stringing of lights on Lions Island signals the start of the Yuletide season.*

▲ *Winter walkers prepare to meet a baby.*

➤ *Parents, children, and pets stroll along the Kessler Boulevard side of the Lake.*

▲ *A sun break allows children to escape from indoors to the Martin's Dock playground.*

◄ *The homes along Nichols Boulevard in the Olympic Addition are mirrored in the Lake.*

Spring Spirits

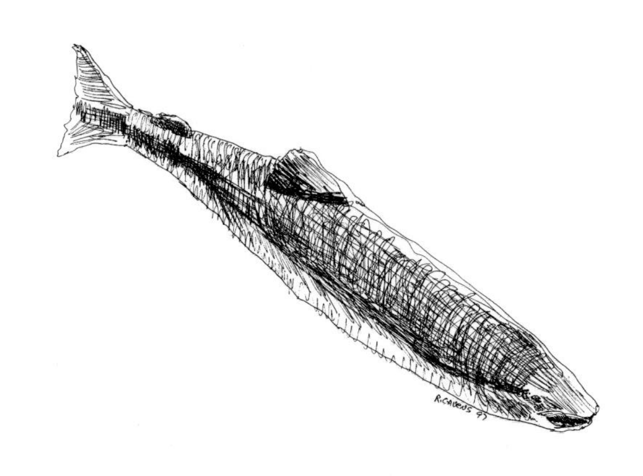

◄ *A good dad shares the excitement of fishing.*

GONE FISHING

They are often seen at the Lake. People, young and old, waiting with frozen body and expression for the faintest nibble on a transparent line trailing in the water. They are fishermen.

And even though they are city fishing, they have high hopes of catching fish and with good cause, for there are fish in the Lake. It has been planted yearly since 1950. Recently, 7,000 brown trout and 16,000 rainbow were added. Furthermore, the fish are good to eat, since the water in the Lake is as clean as any river.

The Cowlitz Game and Anglers Club works with the Park and Recreation Department to put on fishing clinics, where children between the ages of 5 and 12 are taught by the "old timers."

Not all fish in the Lake can be taken. There are signs posted around the Lake with a picture of a particular fish that must be thrown back if caught. This is the grass-eating carp.

For years the grass in the Lake has been a problem, since it grows so heavily and chokes the water, causing it to be stagnate. The parks department had tried using a big orange-colored floating machine that chews up the grass, but it proved to be a losing fight.

There appeared from Arkansas, however, a rescue solution — grass-eating carp. In a seemingly weird turnabout of biology, these fish literally eat the grass. In October, 1994, the city bought 550 of them for $4,200, and they're clearing the Lake. Since the carp are sterile, they will need to be replaced eventually, but should last 7 to 10 years.

◀ *Dawn fishing.*

▶ *A live one is caught.*

▼ *Enthralled children are shown the basics of bait preparation by a member of the Cowlitz Game and Anglers Club at the fishing clinic.*

A radio-controlled model boat is readied for launching.

Wagons west for playtime.

Rain doesn't dampen the interest of children learning to grow a plant at one of the many exhibits of Earth Day.

➤ *Kids sit in wonderment as "Dr. Wilderness" sneaks in an environmental message by the use of magic.*

SEATTLE

◄ *Registration for the Lower Columbia Multiple Sclerosis walkathon.*

► *A group of four among 200 walkers help raise $20,000 for MS.*

▼ *Members of the Cowlitz Valley Runners set the pace in the 5-kilometer race, part of the city-sponsored Earth Day.*

MS-KETEERS
THE MS WALK

GOOSEMAN OF THE LAKE

Tito Vargas was a friend to the geese of the Lake for 20 years. After each day's work as a custodian at Community Church, he would drive his car to the lake and honk his horn, announcing feeding time. The geese would respond with their own wild honking. Vargas would flap his arms in wing-like motions, bend down, and kiss some of them on their bills.

When his wife, Rosa, accompanied him, the geese would bruise her legs by nipping at them in what was thought to be a jealous reaction to her presence. When Tito died in 1989, she paid for a park bench to mark his memory.

A woman in the rain rests on the bench commemorating Tito Vargas.

➤ *Domestic geese of the Lake.*

THE LAKE IS ALIVE!

At early dawn each Christmas Day, some two dozen members of the Willapa Hills Audubon Society, armed only with notepads and binoculars, begin a silent stalk of the birds of the Lake, counting them as they go. There is purpose in what they do, for their results are fed into a national data base that shows changes, up or down, good or bad, of bird populations.

And they have found that Lake Sacajawea Park is alive with a variety of birds, 34 species of water birds and 44 species of land birds.

American coots

WATER BIRDS

Canada goose
Domestic goose
Mallard
Shoveler
Domestic duck
Cormorant
Widgeon
Coot
Ring-billed gull
Herring gull
Glaucous-winged gull
Dusky Canada goose
Green-winged teal
White-winged scoter
Ring-necked duck
Common goldeneye
Bufflehead
Mew gull
Great blue heron
Green-backed heron
Pied-billed grebe
Least grebe
Western gull
Western grebe
Tundra swan
Wood duck
Cinnamon teal
Gaswall
Scaup
Barrow's goldeneye
Ruddy duck
Caspian tern
Osprey

LAND BIRDS

Violet-green swallow
Scrub jay
Crow
Black-capped chickadee
Robin
Starling
Red-winged blackbird
White-crowned sparrow
Song sparrow
Junco
House finch
Pine siskin
Goldfinch
Evening grosbeak
House sparrow
Rock dove (pigeon)
Anna's hummingbird
Rufous hummingbird
Downy woodpecker
Flicker
Barn swallow
Bushtit
House swan
Golden-crowned kinglet
Ruby-crowned kinglet
Cedar waxwing
Yellow-rumped warbler
Rufous-sided towhee
Black-headed grosbeak
Purple finch
Red-tailed hawk
Pileated woodpecker
Hairy woodpecker
Sapsucker
Tree swallow
Stellars jay
Chestnut-back chickadee
Nuthatch
Brown creeper
Bewick's wren
Marsh wren
Warbling vireo
Yellow warbler
Brewer's blackbird

List compiled by Ruth Deery and members of Willapa Hills chapter of the Audubon Society.

Mallards

Squirrels are a common sight at the Lake.

➤ *Riders on horseback are seen occasionally.*

The annual Pet Trek raises money for the Humane Society of Cowlitz County.

➤ *A hiker and her dogs use the underpass of the Louisiana Street bridge.*

KESSLER ELEMENTARY SCHOOL

Kessler School, pictured in 1925, contained grades one to 12.

◄ *Kessler Elementary School.*

THE FIRST SCHOOL

In 1924, Longview grade and high school students all attended classes in one building, Kessler School. But by 1928, the enrollment had grown so much that grades 9 to 12 moved out to the new high school across the Lake, leaving Kessler as the elementary school.

The school was named after George Kessler who drew the initial plans for Longview and Lake Sacajawea Park. He was a nationally known landscape architect, who, in addition to designing the grounds of the St. Louis World's Fair, had done work for Kansas City, Dallas, Wichita, and Oklahoma City.

At the time Kessler School opened in 1924, segregation existed. When a brave Victoria Freeman marched into the classroom with her two black children, Oliver and Calvin, the teacher proclaimed, "Oh no, there's been some mistake" and forbid them to sit down. Mrs. Freeman told her boys to take a seat anyway. The teacher rushed to the office, only to find out that the school principal had been instructed to let the boys remain in class. Mrs. Freeman had already gone to the chairman of the school board to remind him of Washington State law that prohibited segregation in the schools. The boys stayed, and later they graduated from R.A. Long High School with their white classmates.

There is now a new Kessler Elementary School that doesn't resemble the old two-story frame structure in the slightest. In 1974, the original school was torn down, except for the gymnasium, and was rebuilt as a one-story brick building with open classrooms.

THE HOSPITAL BUILDS AND REBUILDS

In 1925 a Long-Bell executive, J.D. Tennant, spurred the drive to construct an 80-bed, 4-story hospital. When it opened as Longview Community Hospital, an experiment in prepaid medicine was started. The hospital contracted with the local industries to provide complete medical care for their employees for $1.00 per person per month.

The hospital began to lose money with this plan, however, and so it started its own outpatient clinic to cut costs, hiring three of the town's doctors. The remaining 12 doctors were allowed to continue admitting patients, but had to guarantee each patient's hospital bill.

This caused a bitter split that went on for decades. The independent doctors kicked the clinic doctors out of the medical society and opened their own hospital, Cowlitz General, in the old train station.

In 1943 Longview Memorial Hospital went broke and closed its doors, until the Sisters of St. Joseph of Newark were persuaded to take on its debt and reopen it. They renamed it St. John's Hospital, and by 1952 were doing well enough to add a 60-bed wing.

In the 1960s, both hospitals competed fiercely for federal dollars to build. Cowlitz got the award and with community funds finished a $2.6 million hospital in April of 1968. But the sisters at St. John's were undaunted. They borrowed the money and were able to dedicate their eight-story $7.8 million addition on May 11, the very next month.

The smaller Cowlitz General Hospital eventually sold out, leaving St. John Medical Center all alone to continue to build.

➤ *The first three stages of St. John Medical Center's building program are shown with the 1924 building on the right, the 1952 addition joining it in the middle, while the 1968 tower rises in the background.*

PeaceHealth

NURTURING THE PARK

The beauty of the Lake did not blossom forth by accident. It was carefully created by acclaimed Kansas City architects George Kessler and S. Robert Hare. Their blueprints, though fading, still show the intended location of every tree and shrub in the Park.

Their original dream continues to grow. The current park superintendent, Al George, envisions the development of an elaborate arboretum, where the trees are all labeled, ready for exhibition and study. Among the 819 trees in the Park there are already 80 different species. Each tree is lovingly listed in a computer where its height, canopy width, and condition are noted.

In addition there are ennobling elms lining Kessler and Nichols boulevards that encircle the Park. But they are in real danger of being wiped out. Dutch elm disease surrounds them, both north and south, in Tacoma, Washington, and Portland, Oregon. This concern has led the park department to begin raising elm saplings that are resistant to the disease.

All of this effort is part of what has been termed urban forestry, the new science of managing timberland amid houses and asphalt. Longview has done well at this, having been named a Tree City, USA, with the value of the city's trees placed at over $53,000,000.

New visual delights are planned for Lake Sacajawea. A Japanese garden is being created on one of its islands, while more lampposts are being erected for a nighttime firefly glow.

➤ *The budding leaves of the surrounding elm trees are a soft pale green in the early spring.*

➤ *Runner Tracy Davis tries to avoid getting wet. This won't be a problem in the future as the sprinkling system will be underground.*

◀ *The Lake is dotted with islands of light, allowing joggers, such as John Crook, to use it day and night.*

▼ *For decades, there was little upkeep of the Park, but now there are funds to maintain its beauty.*

There is full floral bloom on Lions Island in the spring.